# YOUR KNOWLEDGE HAS VALUE

**Bibliographic information published by the German National Library:**

The German National Library lists this publication in the National Bibliography; detailed bibliographic data are available on the Internet at http://dnb.dnb.de .

**Imprint:**

Copyright © 2012 GRIN Verlag, Open Publishing GmbH
Print and binding: Books on Demand GmbH, Norderstedt Germany
ISBN: 9783668325975

**This book at GRIN:**

http://www.grin.com/en/e-book/342263/first-conjecture-on-nonelementary-functions

**Dharmendra Kumar Yadav**

# First Conjecture on Nonelementary Functions

## An Application of strong Liouville's theorem

GRIN Publishing

**GRIN - Your knowledge has value**

Since its foundation in 1998, GRIN has specialized in publishing academic texts by students, college teachers and other academics as e-book and printed book. The website www.grin.com is an ideal platform for presenting term papers, final papers, scientific essays, dissertations and specialist books.

**Visit us on the internet:**

http://www.grin.com/

http://www.facebook.com/grincom

http://www.twitter.com/grin_com

# First Conjecture on Nonelementary Functions

## Dharmendra Kumar Yadav

Assistant Professor, Department of Mathematics

Shivaji College, University of Delhi, Raja Garden, Delhi-27


## Dipak Kumar Sen

Retired Associate Professor & Head, Department of Mathematics

R. S. More College, Govindpur, Dhanbad-09, Jharkhand

**Abstract**

In the paper the proof of one of the Yadav's six conjectures of indefinite nonintegrable functions, classically known as nonelementary functions, and their examples have been discussed by applying *strong Liouville's theorem*, its special case, some well-known nonelementary functions and properties mentioned by *Marchisotto & Zakeri*.


**Key Words:** Nonelementary functions, strong Liouville's theorem etc.


**2010 AMS Subject Classification:** 26A09, 26Bxx

**Introduction**

A natural query arises in integral calculus in indefinite integration that *"what type of functions cannot be integrated?"* or *"which indefinite integrals are not elementary?"* The first example which leads us beyond the region of elementary functions is the *elliptic integrals* due to *John Wallis* (1655). Such integrals cannot be evaluated in terms of the elementary functions was proved by *Joseph Liouville* in 1833 and the main results on functions with nonelementary integrals began with *Strong Liouville Theorem (1835)* and *Strong Liouville Theorem (special case, 1835)*. *Marchisotto & Zakeri* (1994) studied nonelementary functions and mentioned two

1

important examples 4 and 5 [5, pp.300-301], which are treated as properties in proving the functions elementary and nonelementary. By applying the above properties we get the following well-known nonelementary functions:

$$\int e^{x^2}dx \, , \, \int e^{ax^2}dx, (a \neq 0), \, \int e^{-x^2}dx \, , \, \int \frac{e^{-x}}{x}dx \, , \, \int \frac{e^x}{x}dx \, , \, \int \frac{\sin x}{x}dx \, , \, \int \frac{\cos x}{x}dx$$

**First Conjecture on Nonelementary Functions:** *Yadav & Sen* [6, 7] have given six standard forms of indefinite nonintegrable functions out of which form-1 is as follows:

*"An indefinite integral of the form* $\int \dfrac{e^{f(x)}}{f'(x)}dx$ *, where f(x) is a polynomial function of degree* $\geq 2$,

*or a trigonometric (not inverse trigonometric) function, or a hyperbolic (not inverse hyperbolic) function is always nonintegrable i. e. nonelementary."*

**Proof:** We will prove it taking different possible cases as follows:

**Case I:** When f(x) is an algebraic function (polynomial) of degree $\geq 2$:

We have $\int \dfrac{e^{f(x)}}{f'(x)}dx = \int g(x)e^{f(x)}dx$, [Taking $g(x) = \dfrac{1}{f'(x)}$]. From strong Liouville theorem (special case), $\int g(x)e^{f(x)}dx$ is elementary if and only if there exists a rational function R(x) such that

$$g(x) = R'(x) + R(x)f'(x) \implies \frac{1}{f'(x)} = R'(x) + R(x)f'(x)$$

Let $R(x) = \dfrac{p(x)}{q(x)}$, where g.c.d.(p(x), q(x))=1. Then we have

$$\frac{1}{f'(x)} = R'(x) + R(x)f'(x)$$

$$\implies f'(x)q(x)p'(x) - f'(x)p(x)q'(x) + [f'(x)]^2 p(x)q(x) = [q(x)]^2 \quad (1.1)$$

$$\implies f'(x)p'(x) - q(x) + [f'(x)]^2 p(x) = \frac{f'(x)p(x)q'(x)}{q(x)}$$

2

Which implies $q(x)|f'(x)$ as $q(x)$ cannot divide $p(x)$ and $q'(x)$. In this case either $q(x)=k$, a constant or a polynomial of degree less than or equal to the degree of $f'(x)$.

For $q(x)=k$, from (1.1) we have $f'(x)kp'(x)+[f'(x)]^2 p(x)k = k^2$        (1.2)

Comparing degrees of x in (1.2) results out in a contradiction. Hence $q(x)$ cannot be a constant.

For $q(x)$ a polynomial of degree less than or equal to the degree of $f'(x)$, we have since $q(x)|f'(x)$, let us assume that $f'(x)=q(x).h(x)$. Then from (1.1)

$$q(x)h(x)q(x)p'(x) - q(x)h(x)p(x)q'(x) + [q(x)h(x)]^2 p(x)q(x) = [q(x)]^2$$

$$\Rightarrow h(x)p'(x) - 1 + q(x)[h(x)]^2 p(x) = \frac{h(x)p(x)q'(x)}{q(x)}$$        (1.3)

Which implies $q(x)|h(x)$, since $q(x)$ cannot divide $p(x)$ and $q'(x)$. Let $h(x)=q(x).\xi(x)$. Then from (1.3), we have

$$q(x)\xi(x)p'(x) - p(x)q'(x)\xi(x) + [q(x)]^3 [\xi(x)]^2 p(x) = 1$$        (1.4)

Comparing the degrees of x in both sides in (1.4) results out in a contradiction. Therefore such $R(x)$ does not exist, i. e., the given function is nonelementary.

**Case II: When f(x) be a trigonometric (not inverse trigonometric) function:**

Let us consider them one by one.

**1.1** For sine function, we have

$$\int \frac{e^{f(x)}dx}{f'(x)} = \int \frac{e^{\sin \varphi(x)}dx}{\varphi'(x)\cos \varphi(x)},$$

where $\varphi(x)$ be any polynomial of degree $\geq 1$. On putting $\sin\varphi(x)=z$, we have

$$\int \frac{e^{\sin\varphi(x)}dx}{\varphi'(x)\cos \varphi(x)} = \int \frac{e^z dz}{[\varphi'(x)]^2 (1-z^2)}$$        (1.1.1)

**Sub-case I:** When $\varphi(x)$ is linear, let $\varphi(x)=x+b$. Then from (1.1.1) we have

$$\int \frac{e^z dz}{[\varphi'(x)]^2 (1-z^2)} = \int \frac{e^z dz}{(1-z^2)} = \frac{1}{2}\left[ \int \frac{e^z dz}{(1-z)} + \int \frac{e^z dz}{(1+z)} \right]$$

3

where

$$\int \frac{e^z dz}{(1-z)} = -e\int \frac{e^{-p}}{p} dp, [\text{Putting } 1 - z = p]$$

and
$$\int \frac{e^z dz}{(1+z)} = \frac{1}{e}\int \frac{e^p}{p} dp, [\text{Putting } 1 + z = p]$$

Both are nonelementary from example-4 due to Marchisotto et al [5, pp.300]. Therefore the given function is also nonelementary.


**Sub-case II**: When $\varphi(x)$ is a polynomial of degree 2. Let us consider $\varphi(x)=x^2+bx+c$. Then we have from (1.1.1), on putting $\sin\varphi(x)=z$

$$\int \frac{e^z dz}{[\varphi'(x)]^2(1-z^2)} = \frac{1}{4}\int \frac{e^z dz}{[\sin^{-1}z+k](1-z^2)}, \text{where } k = \frac{b^2-4c}{4}$$

$$= \frac{1}{4}\int \frac{e^z dz}{[\sin^{-1}z+k]\sqrt{(1-z^2)}\sqrt{(1-z^2)}}$$

$$= \int F\left[z, e^z, \sqrt{1-z^2}, \sin^{-1}z\right] dz = \int F[z, y_1, y_2, y_3] dz$$

$$\left[\frac{dy_1}{dz} = e^z = y_1, \frac{dy_2}{dz} = \frac{-z}{\sqrt{1-z^2}} = \frac{-z}{y_2}, \frac{dy_3}{dz} = \frac{1}{\sqrt{1-z^2}} = \frac{1}{y_2}\right]$$

Applying strong Liouville theorem, part (b), we find that it is elementary if and only if there exists an identity of the form

$$\frac{e^z}{4[\sin^{-1}z+k](1-z^2)} = \frac{d}{dz}\left[U_0 + \sum_{i=1}^{n} c_i \log U_i\right]$$

$$\Rightarrow \frac{e^z}{4[\sin^{-1}z+k](1-z^2)} = \left[\frac{dU_0}{dz} + \sum_{i=1}^{n} c_i \frac{U'_i}{U_i}\right]$$

where each $U_j$ is a function of z, $y_1$, $y_2$, and $y_3$. Considering different forms of $U_j$ like

$$\log[(\sin^{-1}z+k)e^z], e^z \log(\sin^{-1}z+k)$$

we find that no such $U_j$ exist. Hence the given function is nonelementary. Similarly we can prove it nonelementary for higher degree polynomials $\varphi(x)$.

4

**1.2** For cosine function, we have

$$\int \frac{e^{f(x)}dx}{f'(x)} = \int \frac{e^{\cos\varphi(x)}dx}{-\varphi'(x)\sin\varphi(x)}$$

where $\varphi(x)$ be any polynomial of degree $\geq 1$. On putting $\cos\varphi(x)=z$, we have

$$\int \frac{e^{\cos\varphi(x)}dx}{-\varphi'(x)\sin\varphi(x)} = \int \frac{e^z dz}{[-\varphi'(x)]^2(1-z^2)} = \int \frac{e^z dz}{[\varphi'(x)]^2(1-z^2)} \qquad (1.2.1)$$

**Sub-case I:** When $\varphi(x)$ is linear, let $\varphi(x)=x+b$. Then from (1.2.1) we have

$$\int \frac{e^z dz}{[\varphi'(x)]^2(1-z^2)} = \int \frac{e^z dz}{(1-z^2)}$$

which is nonelementary proved in section 1.1 subcase-I.

**Sub-case II:** When $\varphi(x)$ is a polynomial of degree 2. Then from (1.2.1) we have

$$\int \frac{e^z dz}{[\varphi'(x)]^2(1-z^2)} = \frac{1}{4}\int \frac{e^z dz}{(\cos^{-1}z+k)(1-z^2)}$$

$$= \int F[z,e^z,\sqrt{1-z^2},\cos^{-1}z]dz$$

A similar argument will hold as in section 1.1 to prove it nonelementary.

**1.3.** For tangent function, we have on putting $\tan\varphi(x)=z$.

$$\int \frac{e^{f(x)}dx}{f'(x)} = \int \frac{e^{\tan\varphi(x)}dx}{\varphi'(x)\sec^2\varphi(x)} = \int \frac{e^z dz}{[\varphi'(x)]^2(1+z^2)} \qquad (1.3.1)$$

**Sub-case I:** When $\varphi(x)$ is linear, let $\varphi(x)=x+b$. Then from (1.3.1) we have

$$\int \frac{e^z dz}{[\varphi'(x)]^2(1+z^2)} = \int \frac{e^z dz}{(1+z^2)} = \frac{1}{2}\left[\int \frac{e^z dz}{(1+iz)} + \int \frac{e^z dz}{(1-iz)}\right]$$

Now

$$\int \frac{e^z dz}{(1+iz)} = \frac{e^i}{i}\int \frac{e^{-ip}}{p}dp, \text{ putting } (1+iz)=p$$

5

By strong Liouville theorem (special case), it is elementary if and only if there exists a rational function R(x) such that it satisfies the identity

$$\frac{1}{p} = R'(p) - iR(p) \Rightarrow R(p) = 0 \text{ and } R'(p) = \frac{1}{p}$$

But R(p) cannot be zero, so such R(p) does not exist. Hence it is nonelementary.

Also

$$\int \frac{e^z dz}{(1-iz)} = ie^{-i}\int \frac{e^{ip}}{p} dp, \text{ putting } (1-iz) = p$$

Again by strong Liouville theorem (special case), it is elementary if and only if there exists a rational function R(x) which satisfies the identity

$$\frac{1}{p} = R'(p) + iR(p) \Rightarrow R(p) = 0 \text{ and } R'(p) = \frac{1}{p}$$

But R(p) cannot be zero, so such R(p) does not exist. Hence it is nonelementary. Therefore the given function is nonelementary in this case.


**Sub-case II**: When $\varphi(x)=x^2+bx+c$. Then from (1.3.1) we have

$$\int \frac{e^z dz}{[\varphi'(x)]^2 (1+z^2)} = \frac{1}{4}\int \frac{e^z dz}{[\tan^{-1}z+k](1+z^2)}, k = \frac{b^2-4c}{4}$$

$$= \int F[z, e^z, (1+z^2), \tan^{-1}z] dz = \int F[z, y_1, y_2, y_3] dz$$

$$\left[ \frac{dy_1}{dz} = e^z = y_1, \frac{dy_2}{dz} = 2z, \frac{dy_3}{dz} = \frac{1}{1+z^2} = \frac{1}{y_2} \right]$$

By strong Liouville theorem part (b), it is elementary if and only if there exists an identity of the form containing $U_j$, a function of z, $y_1$, $y_2$, and $y_3$

$$\frac{e^z}{4[\tan^{-1}z+k](1+z^2)} = \frac{dU_i}{dz} + \sum_{i=1}^{n} c_i \frac{U_i'}{U_i}$$

Considering different forms of $U_j$ like $e^z \log(\tan^{-1}z+k)$, $\log[e^z(\tan^{-1}z+k)]$, etc. we find that no such $U_j$ exist, i. e., no such identity exist. Hence the given function is nonelementary. Similarly we can prove it nonelementary for higher degree polynomials $\varphi(x)$.

6

**1.4.** For cotangent function, we have on putting $\cot\varphi(x)=z$

$$\int\frac{e^{f(x)}dx}{f'(x)}=\int\frac{e^{\cot\varphi(x)}dx}{-\varphi'(x)\cos ec^2\varphi(x)}=\int\frac{e^z dz}{[\varphi'(x)]^2(1+z^2)}$$

(1.4.1)

**Sub-case-I**: For $\varphi(x)=x+b$, we have from (1.4.1)

$$\int\frac{e^z dz}{[\varphi'(x)]^2(1+z^2)}=\int\frac{e^z dz}{(1+z^2)}$$

Which is nonelementary, proved in section 1.3, sub-case-I.

**Sub-case-II**: For $\varphi(x)=x^2+bx+c$, we have from (1.4.1)

$$\int\frac{e^z dz}{[\varphi'(x)]^2(1+z^2)}=\frac{1}{4}\int\frac{e^z dz}{(\cot^{-1}z+k)(1+z^2)},k=\frac{b^2-4c}{4}$$

$$=\int F[z,e^z,(1+z^2),\cot^{-1}z]dz$$

A similar argument will hold as in section 1.3 to prove it nonelementary.

**1.5.** For cosecant function, we have on putting $\csc\varphi(x)=z$

$$\int\frac{e^{f(x)}dx}{f'(x)}=\int\frac{e^{\csc\varphi(x)}dx}{-\varphi'(x)\cos ec\varphi(x)\cot\varphi(x)}=\int\frac{e^z dz}{[\varphi'(x)]^2 z^2(z^2-1)}$$

(1.5.1)

**Sub-case I**: When $\varphi(x)$ is linear, let $\varphi(x)=x+b$. Then from (1.5.1) we have

$$\int\frac{e^z dz}{[\varphi'(x)]^2 z^2(z^2-1)}=\int\frac{e^z dz}{z^2(z^2-1)}=\left[\int\frac{e^z dz}{(z^2-1)}-\int\frac{e^z dz}{z^2}\right]$$

Where the first integral

$$\int\frac{e^z dz}{(z^2-1)}$$

is nonelementary as proved in section 1.1, sub-case-I and the second integral

$$\int\frac{e^z dz}{z^2}=\int z^{-2}e^z dz$$

is also nonelementary from example-5 due to Marchisotto et al [5, pp.301].

7

**Sub-case II**: When $\varphi(x)=x^2+bx+c$, then from (1.5.1) we have

$$\int \frac{e^z dz}{[\varphi'(x)]^2 z^2 (z^2-1)} = \int \frac{e^z dz}{[2x+b]^2 z^2 (z^2-1)}$$

$$= \int \frac{e^z dz}{4[\cos ec^{-1}z+k]z^2(z^2-1)}, k = \frac{b^2-4c}{4}$$

$$= \int F[z,e^z,\sqrt{z^2-1},\cos ec^{-1}z]dz = \int F[z,y_1,y_2,y_3]dz$$

$$\left[\frac{dy_1}{dz}=e^z=y_1, \frac{dy_2}{dz}=\frac{z}{\sqrt{z^2-1}}=\frac{z}{y_2}, \frac{dy_3}{dz}=\frac{-1}{|z|\sqrt{z^2-1}}=\frac{-1}{|z|y_2}\right]$$

By strong Liouville theorem part (b), this is elementary if and only if there exists an identity of the form containing $U_j$, a function of z, $y_1$, $y_2$, and $y_3$ as follows

$$\frac{e^z}{4[\cos ec^{-1}z+k]z^2(z^2-1)} = \frac{dU_i}{dz} + \sum_{i=1}^{n} c_i \frac{U_i'}{U_i}$$

Considering different forms of $U_j$ like $e^z\log[\cosec^{-1}z+k]$, $\log[e^z(\cosec^{-1}z+k)]$, etc., we find that no such $U_j$ exist, which satisfy the above identity. Hence the given function is nonelementary. Similarly we can prove it for higher degree polynomial $\varphi(x)$.

**1.6.** For secant function, we have on putting $\sec\varphi(x)=z$

$$\int \frac{e^{f(x)}dx}{f'(x)} = \int \frac{e^{\sec\varphi(x)}dx}{\varphi'(x)\sec\varphi(x)\tan\varphi(x)} = \int \frac{e^z dz}{[\varphi'(x)]^2 z^2(z^2-1)} \qquad (1.6.1)$$

**Sub-case-I**: For $\varphi(x)=x+b$, we have from (1.6.1)

$$\int \frac{e^z dz}{[\varphi'(x)]^2 z^2(z^2-1)} = \int \frac{e^z dz}{z^2(z^2-1)}$$

Which is nonelementary, proved in section 1.5, sub-case-I.

**Sub-case-II**: For $\varphi(x)=x^2+bx+c$, we have from (1.6.1)

$$\int \frac{e^z dz}{[\varphi'(x)]^2 z^2 (z^2-1)} = \frac{1}{4}\int \frac{e^z dz}{[\sec^{-1} z + k]z^2 (z^2-1)}, k = \frac{b^2 - 4c}{4}$$

$$= \int F[z, e^z, \sqrt{z^2-1}, \sec^{-1} z]dz$$

It can be proved nonelementary by the similar procedure as has been done in section 1.5.

**Case III: When f(x) be a hyperbolic (not inverse hyperbolic) function.** Let us consider them one by one.

**1.7.** For sine hyperbolic function, we have on putting $\sinh\varphi(x)=z$

$$\int \frac{e^{f(x)}dx}{f'(x)} = \int \frac{e^{\sinh\varphi(x)}dx}{\varphi'(x)\cosh\varphi(x)} = \int \frac{e^z dz}{[\varphi'(x)]^2(1+z^2)} \tag{1.7.1}$$

**Sub-case I:** When $\varphi(x)$ is linear, let $\varphi(x)=x+b$. Then from (1.7.1) we have

$$\int \frac{e^z dz}{[\varphi'(x)]^2(1+z^2)} = \int \frac{e^z dz}{(1+z^2)}$$

which is nonelementary, proved in section 1.3, sub-case-I.

**Sub-case II:** When $\varphi(x)=x^2+bx+c$. Then from (1.7.1), we have

$$\int \frac{e^z dz}{[\varphi'(x)]^2(1+z^2)} = \int \frac{e^z dz}{[2x+b]^2(1+z^2)}$$

$$= \int \frac{e^z dz}{4(\sinh^{-1} z+k)(1+z^2)}, \text{ where } k = \frac{b^2-4c}{4}$$

$$= \int F[z, e^z, \sqrt{1+z^2}, \sinh^{-1} z]dz = \int F[z, y_1, y_2, y_3]dz$$

$$\left[\frac{dy_1}{dz} = e^z = y_1, \frac{dy_2}{dz} = \frac{z}{\sqrt{1+z^2}}, \frac{z}{y_2}, \frac{dy_3}{dz} = \frac{1}{\sqrt{1+z^2}} = \frac{1}{y_2}\right]$$

Applying strong Liouville theorem part (b), it is elementary if and only if there exists an identity of the form

9

$$\frac{e^z}{4(\sinh^{-1}z+k)(1+z^2)} = \frac{dU_0}{dz} + \sum_{i=1}^{n} c_i \frac{U_i'}{U_i}.$$

Considering different possible forms of $U_j$ like $e^z\log[\sinh^{-1}z+k]$, $\log[e^z(\sinh^{-1}z+k)]$, etc. we find that no such $U_j$ exist. Hence the given function is nonelementary. Similarly we can prove it nonelementary for higher degree polynomials.

**1.8.** For cosine hyperbolic function, we have on putting $\cosh\varphi(x)=z$

$$\int\frac{e^{f(x)}dx}{f'(x)} = \int\frac{e^{\cosh\varphi(x)}dx}{\varphi'(x)\sinh\varphi(x)} = \int\frac{e^z dz}{[\varphi'(x)]^2(z^2-1)} \qquad (1.8.1)$$

**Sub-case-I**: For $\varphi(x)=x+b$, we have from (1.8.1)

$$\int\frac{e^z dz}{[\varphi'(x)]^2(z^2-1)} = \int\frac{e^z dz}{(z^2-1)}$$

Which is nonelementary proved in section 1.1, sub-case-I.

**Sub-case-II**: For $\varphi(x)=x^2+bx+c$, we have from (1.8.1)

$$\int\frac{e^z dz}{[\varphi'(x)]^2(z^2-1)} = \frac{1}{4}\int\frac{e^z dz}{(\cosh^{-1}z+k)(z^2-1)}, k=\frac{b^2-4c}{4}$$

$$= \int F[z,e^z,\sqrt{z^2-1},\cosh^{-1}z]dz$$

It can now be proved nonelementary by strong Liouville theorem part (b). Similarly we can prove it for higher degree polynomial $\varphi(x)$.

**1.9.** For tangent hyperbolic function, we have on putting $\tanh\varphi(x)=z$

$$\int\frac{e^{f(x)}dx}{f'(x)} = \int\frac{e^{\tanh\varphi(x)}dx}{\varphi'(x)\sec h^2\varphi(x)} = \int\frac{e^z dz}{[\varphi'(x)]^2(1-z^2)} \qquad (1.9.1)$$

**Sub-case-I**: For $\varphi(x)=x+b$, we have from (1.9.1)

$$\int \frac{e^z dz}{[\varphi'(x)]^2(1-z^2)} = \int \frac{e^z dz}{(1-z^2)}$$

Which is nonelementary, proved in section 1.1, sub-case-I.


**Sub-case-II**: For $\varphi(x)=x^2+bx+c$, we have from (1.9.1)

$$\int \frac{e^z dz}{[\varphi'(x)]^2(1-z^2)} = \frac{1}{4}\int \frac{e^z dz}{(\tanh^{-1} z+k)(1-z^2)}, \quad k=\frac{b^2-4c}{4}$$

$$= \int F[z,e^z,(1-z^2),\tanh^{-1} z]dz$$

It can now be proved nonelementary by strong Liouville theorem part (b). Similarly we can prove it for higher degree polynomial $\varphi(x)$.


**1.10.** For cotangent hyperbolic function, we have on putting $\coth\varphi(x)=z$

$$\int \frac{e^{f(x)}dx}{f'(x)} = \int \frac{e^{\cot\varphi h(x)}dx}{-\varphi'(x)\cos ech^2\varphi(x)} = \int \frac{e^z dz}{[\varphi'(x)]^2(z^2-1)} \qquad (1.10.1)$$


**Sub-case-I**: For $\varphi(x)=x+b$, we have from (1.10.1)

$$\int \frac{e^z dz}{[\varphi'(x)]^2(z^2-1)} = \int \frac{e^z dz}{(z^2-1)}$$

Which is nonelementary proved in section 1.1, sub-case-I.


**Sub-case-II**: For $\varphi(x)=x^2+bx+c$, we have from (1.10.1)

$$\int \frac{e^z dz}{[\varphi'(x)]^2(z^2-1)} = \frac{1}{4}\int \frac{e^z dz}{(\coth^{-1} z+k)(z^2-1)}, \quad k=\frac{b^2-4c}{4}$$

$$= \int F[z,e^z,(z^2-1),\coth^{-1} z]dz$$

It can now be proved nonelementary by strong Liouville theorem part (b). Similarly we can prove it for higher degree polynomial $\varphi(x)$.


**1.11.** For cosecant hyperbolic function, we have on putting $\text{cosech}\varphi(x)=z$

$$\int \frac{e^{f(x)}dx}{f'(x)} = \int \frac{e^{\cos ec\varphi h(x)}dx}{-\varphi'(x)\cos ech\varphi(x)\coth\varphi(x)} = \int \frac{e^{z}dz}{[\varphi'(x)]^{2}z^{2}(z^{2}+1)} \qquad (1.11.1)$$

**Sub-case-I**: For $\varphi(x)=x+b$, we have from (1.11.1)

$$\int \frac{e^{z}dz}{[\varphi'(x)]^{2}z^{2}(z^{2}+1)} = \int \frac{e^{z}dz}{z^{2}(z^{2}+1)} = \int \frac{e^{z}dz}{z^{2}} - \int \frac{e^{z}dz}{z^{2}+1}$$

Both are nonelementary proved in section 1.5, sub-case-I and section 1.7, sub-case-I respectively.

**Sub-case-II**: For $\varphi(x)=x^{2}+bx+c$, we have from (1.11.1)

$$\int \frac{e^{z}dz}{[\varphi'(x)]^{2}z^{2}(z^{2}+1)} = \frac{1}{4}\int \frac{e^{z}dz}{(\cos ech^{-1}z+k)z^{2}(z^{2}+1)}, k = \frac{b^{2}-4c}{4}$$

$$= \int F[z,e^{z},\sqrt{z^{2}+1},\cos ech^{-1}z]dz$$

It can now be proved nonelementary by strong Liouville theorem part (b). Similarly we can prove it for higher degree polynomial $\varphi(x)$.

**1.12**. For secant hyperbolic function, we have on putting $sech\varphi(x)=z$

$$\int \frac{e^{f(x)}dx}{f'(x)} = \int \frac{e^{\sec h\varphi(x)}dx}{-\varphi'(x)\sec h\varphi(x)\tanh\varphi(x)} = \int \frac{e^{z}dz}{[\varphi'(x)]^{2}z^{2}(1-z^{2})} \qquad (1.12.1)$$

**Sub-case-I**: For $\varphi(x)=x+b$, we have from (1.12.1)

$$\int \frac{e^{z}dz}{[\varphi'(x)]^{2}z^{2}(1-z^{2})} = \int \frac{e^{z}dz}{z^{2}(1-z^{2})} = \int \frac{e^{z}dz}{z^{2}} + \int \frac{e^{z}dz}{1-z^{2}}$$

Both are nonelementary proved in section 1.5, sub-case-I and section 1.1, sub-case-I respectively.

**Sub-case-II**: For $\varphi(x)=x^{2}+bx+c$, we have from (1.12.1)

$$\int \frac{e^z dz}{[\varphi'(x)]^2 z^2 (1-z^2)} = \frac{1}{4} \int \frac{e^z dz}{(\sec h^{-1} z + k) z^2 (1-z^2)}$$

$$= \int F[z, e^z, \sqrt{1-z^2}, \sec h^{-1} z] dz$$

It can now be proved nonelementary by strong Liouville theorem part (b). Similarly we can prove it for higher degree polynomial $\varphi(x)$.

Let us consider some examples on this standard form of nonelementary functions:

**Example 1**: Show that the integral $\int \frac{e^{ax^2+b}}{x} dx$, $a \neq 0$ is nonelementary.

Proof: We have

$$\int \frac{e^{ax^2+b}}{x} dx = \int \frac{e^{ax^2}}{x} dx + \int \frac{e^b}{x} dx = e^b \log x + \int \frac{2axe^{ax^2}}{2ax^2} dx$$

Now for second integral, putting $ax^2 = z$ we have

$$\int \frac{2axe^{ax^2}}{2ax^2} dx = \frac{1}{2} \int \frac{e^z}{z} dz = \frac{1}{2} \int z^{-1} e^z dz$$

which is nonelementary from example-5 due to Marchisotto et al [5, pp.301].

**Example 2**: Show that the integral $\int \frac{e^{\sin x}}{\cos x} dx$ is nonelementary.

Proof: We have $\int \frac{e^{\sin x}}{\cos x} dx = \int \frac{e^{\sin x} \cos x}{\cos^2 x} dx = \int \frac{e^z dz}{(1-z^2)}$

On putting $\sin x = z$. Which is nonelementary, proved in section 1.1, sub-case-I.

**Example 3**: Show that the integral $\int \frac{e^{\cos x}}{-\sin x} dx$ is nonelementary.

Proof: We have

$$\int \frac{e^{\cos x}}{-\sin x} dx = \int \frac{e^{\cos x}(-\sin x)}{(-\sin x)^2} dx = \int \frac{e^z dz}{(1-z^2)}$$

On putting $\cos x = z$. Which is nonelementary proved in section 1.1, sub-case-I.

**Example 4:** Show that the integral $\int \dfrac{e^{\tan x}}{\sec^2 x}\,dx$ is nonelementary.

Proof: We have, on putting tanx=z

$$\int \frac{e^{\tan x}}{\sec^2 x}\,dx = \int \frac{e^z}{(1+z^2)^2}\,dz = \frac{1}{4}\int \frac{e^z dz}{(iz)(1-iz)^2} - \frac{1}{4}\int \frac{e^z dz}{(iz)(1+iz)^2} \tag{A}$$

We have on putting 1-iz = p in the first integral of (A)

$$\int \frac{e^z dz}{(iz)(1-iz)^2} = ie^{-i}\left[\int \frac{e^{ip}dp}{(1-p)} + \int \frac{e^{ip}dp}{p} + \int \frac{e^{ip}dp}{p^2}\right] \tag{B}$$

where the second and third integrals are nonelementary from example-5 due to Marchisotto et al [5, pp.301]. Now putting 1-p=X in the first integral of (B) we have

$$\int \frac{e^{ip}dp}{(1-p)} = -e^i\int \frac{e^{-iX}dX}{X}$$

which is also nonelementary from example-5 due to Marchisotto et al [5, pp.301]. Therefore the first integral of (A) is nonelementary. Similarly we can prove that the second integral of (A) is also nonelementary. Therefore the given function is nonelementary.

**Example 5:** Show that the integral $\int \dfrac{e^{\sinh x}}{\cosh x}\,dx$ is nonelementary.

Proof: We have on putting sinhx=z

$$\int \frac{e^{\sinh x}}{\cosh x}\,dx = \int \frac{e^z}{(1+z^2)}\,dz$$

Which is nonelementary proved in section 1.3, sub-case-I.

**Example 6:** Show that the integral $\int \dfrac{e^{\cot x}}{-\cos ec^2 x}\,dx$ is nonelementary.

Proof: We have on putting z=cotx

$$\int \frac{e^{\cot x}}{-\cos ec^2 x}\,dx = \int \frac{e^z}{(1+z^2)}\,dz$$

Which is nonelementary proved in section 1.3, sub-case-I.

**Example 7**: Show that the integral $\int \dfrac{e^{\sec x}}{\sec x.\tan x}\,dx$ is nonelementary.

Proof: We have on putting $\sec x = z$

$$\int \frac{e^{\sec x}}{\sec x.\tan x}\,dx = \int \frac{e^z}{z^2(z^2-1)}\,dz = \int \frac{e^z}{(z^2-1)}\,dz - \int \frac{e^z}{z^2}\,dz$$

Which are nonelementary, proved in section 1.5, sub-case-I.

**Example 8**: Show that the integral $\int \dfrac{e^{\cos ecx}}{-\cos ecx.\cot x}\,dx$ is nonelementary.

Proof: We have on putting $\cos ecx = z$,

$$\int \frac{e^{\cos ecx}}{-\cos ecx.\cot x}\,dx = \int \frac{e^z}{z^2(z^2-1)}\,dz = \int \frac{e^z}{(z^2-1)}\,dz - \int \frac{e^z}{z^2}\,dz$$

Which is nonelementary, proved in section 1.5, sub-case-I.

**Example 9**: Show that the integral $\int \dfrac{e^{\sin^2 x}}{\sin 2x}\,dx$ is nonelementary.

Proof: We have on putting $\sin^2 x = z$,

$$\int \frac{e^{\sin^2 x}}{\sin 2x}\,dx = \frac{1}{4}\int \frac{e^z dz}{z(1-z^2)} = \frac{1}{4}\left[\int \frac{ze^z dz}{(1-z^2)} + \int \frac{e^z dz}{z}\right]$$

Where $\int \dfrac{e^z}{z}\,dz$ is nonelementary from example-5 due to Marchisotto et al [5, pp.301].

Now since $I = \int \dfrac{ze^z}{(1-z^2)}\,dz = \dfrac{1}{2}\int \dfrac{e^z dz}{(1-z)} - \dfrac{1}{2}\int \dfrac{e^z dz}{(1+z)}$

Where $\int \dfrac{e^z dz}{(1-z)} = -e\int \dfrac{e^{-p}}{p}\,dp$, on putting $1-z=p$, which is nonelementary

and

$\int \dfrac{e^z dz}{(1+z)} = \dfrac{1}{e}\int \dfrac{e^p}{p}\,dp$ on putting $1+z=p$, which is also nonelementary

from example-5 due to Marchisotto et al [5, pp.301]. Hence the given function is nonelementary.

**Example 10:** Show that the integral $\int \dfrac{e^{\sin x^2}}{2x.\cos x^2}\,dx$ is nonelementary.

Proof: We have on putting $\sin x^2 = z$,

$$\int \frac{e^{\sin x^2}}{2x.\cos x^2}\,dx = \int \frac{e^z dz}{4(1-z^2)\sin^{-1} z}$$

$$= \int F\!\left(z, e^z, \sqrt{1-z^2}, \sin^{-1} z\right) dz = \int F\!\left(z, y_1, y_2, y_3\right) dz$$

$$\left[ \frac{dy_1}{dz} = e^z = y_1, \frac{dy_2}{dz} = \frac{-z}{\sqrt{1-z^2}} = \frac{-z}{y_2}, \frac{dy_3}{dz} = \frac{1}{\sqrt{1-z^2}} = \frac{1}{y_2} \right]$$

Applying strong Liouville theorem, part(b), it is elementary if and only if there exists an identity of the form, containing $U_i$ a function of $z$, $y_1$, $y_2$, and $y_3$ as

$$\frac{dU_o}{dz} + \sum_{i=1}^{n} C_i \frac{U'_i}{U_i} = \frac{e^z}{(1-z^2)\sin^{-1} z}$$

Taking different possible forms of $U_j$ we find that no such $U_j$ exist. Hence the given function is nonelementary.

**Acknowledgement**

The conjecture discussed in the paper is the first standard form of indefinite nonintegrable functions discussed in chapter two in the doctorate thesis of first author cited in reference [7], which was submitted in the University Department of Mathematics, Vinoba Bhave University, Hazaribag, Jharkhand, in 2012.

**References**

1. Hardy G. H., The Integration of Functions of a Single Variable, 2nd Ed., Cambridge University Press, London, Reprint 1928, **1916**

2. Ritt J. F., Integration in Finite Terms: Liouville's Theory of Elementary Methods, Columbia University Press, New York, **1948**

3. Risch R. H., The Problem of Integration in Finite Terms, Transactions of the American Mathematical Society, 139, 167-189, **1969**

4. Rosenlicht M., Integration in Finite Terms, The American Mathematical Monthly, 79:9, 963-972, **1972**

5. Marchisotto E. A. & Zakeri G. A., An Invitation to Integration in Finite Terms, The College Mathematics Journal, Mathematical Association of America, 25:4, 295- 308, **1994**

6. Yadav D. K. & Sen D. K., Revised paper on Indefinite Nonintegrable Functions, Acta Ciencia Indica, 34:3, 1383-1384, **2008**

7. Yadav D. K., A Study of Indefinite Nonintegrable Functions, Ph. D. Thesis, Vinoba Bhave University, Hazaribag, Jharkhand, **2012, Online:** GRIN Verlag, Munich, Germany, ISBN: 9783668312784, www.grin.com/ebook/341510/

# YOUR KNOWLEDGE HAS VALUE